MACHIAN MOND

MODIFICATION OF NEWTONIAN DYNAMICS

BY MACH'S INERTIA PRINCIPLE

By

MOHSEN LUTEPHY[1]

https://www.researchgate.net/profile/Mohsen_Lutephy

Copyright © 2013

---

[1] Email me at: lutephy@gmail.com

The science needs to be continued truly on the freedom and limitation to the boundaries like the Einstein field equation and playing with it as a simulation leads us to a dark box.

ISBN: 1519144024

ISBN-13: 978-1519144027

INTRODUCTION

Albert Einstein based the theory of relativity regarding to Mach's mechanics (1960) as noted by Mach that

"No one is competent to predicate things about facts. absolute space and absolute motion; they are pure things of thought, pure mental constructs, that cannot be produced in experience. All our principles of mechanics are, as we have shown in detail, experimental knowledge concerning the relative positions and motions of bodies."

Then in Mach's mechanics, the universe should be interconnected as noted by Mach that

"But we must not forget that all things in the world are connected with one another and depend on one another, and that we ourselves and all our thoughts are also a part of nature. It is utterly beyond our power to measure the tune changes of things by time. Quite the contrary, time is an abstraction, at which we arrive by means of the changes of things; made because we are not restricted to any one definite measure, all being interconnected."

For a list of Machian laws we refer to (Bondi and Samuel, 1996) which has referred directly to the book" Mach's Principle from Newton's Bucket to Quantum Gravity" (Barbour and Pfister, 1995) so that

Mach0: The universe, as represented by the average motion of distant galaxies, does not appear to rotate relative to local inertial frames.

c

Mach1: Newton's gravitational constant G is a dynamical field.

Mach2: An isolated body in otherwise empty space has no inertia.

Mach3: Local inertial frames are affected by the cosmic motion and distribution of matter.

Mach4: The universe is spatially closed.

Mach5: The total energy, angular and linear momentum of the universe are zero.

Mach6: Inertial mass is affected by the global distribution of matter.

Mach7: If you take away all matter, there is no more space.

Mach8: $\Omega = 4\pi G\rho T$ is a definite number of order unity.(ρ is universe mean density and T is Hubble time.)

Mach9: The theory contains no absolute elements.

Mach10: Overall rigid rotations and translations of a system are unobservable.

In Newtonian mechanics, the inertia is an intrinsic property of the matter but in the Mach's mechanics, the inertia is perfectly extrinsic as noted by M. Sachs (2003) that

" The latter is the assertion that only the distant stars of the universe determine the mass of any local matter. In contrast to this, in his Science of Mechanics (1883), Mach said that all of the matter of the universe, not only the distant stars, determines the inertial mass of any localized matter. . . .Nevertheless, it was Mach's contention that in principle all of the matter of the closed

system—*the nearby as well as far away constituents—determines the inertial mass of any local matter.*"

Mach realized that the Newtonian mechanics requires to be modified to obtain fully relational dynamics (mere ordering upon actual objects) but he was not success to achieve it. Einstein (1912) with a preliminary scalar theory of gravitation resulted that the presence of the spherical shell of mass M and radius R, increases the inertial mass *m* of the point at its centre as

$$GM_u / c^2 R_u = 1 \tag{1}$$

In this equation when we consider M_u as the mass of universe and R_u as the radius of universe it is extracted so called Machian relation (Einstein-Whitrow-Randall-Sciama-Brans-Dicke relation) which is the mathematical context of the Mach's inertia principle also extracted by different ways by different scientists.

Then Einstein introduces Mach's Principle (1918) "*. . . the presence of the inertial shell K increases the inertial mass of the material point P within it. This makes it plausible that the entire inertia of a mass point is the effect of the presence of all other masses, resulting from a kind of interaction with them. This is exactly the standpoint for which E. Mach has argued persuasively in his penetrating investigations of this matter*".

Also Einstein considered a relativistic version of Mach's inertia principle (the first time this term entered the literature): . . . that the $g_{\mu\nu}$ are completely determined by the mass of bodies, more generally by $T_{\mu\nu}$."

But ultimately Einstein general relativity failed to agree with Mach's inertia principle and Einstein's equations (Einstein, 1959) resulted just a partial dependency.

Several scientists tried to reformulate the Einstein total field to obtain a kind of equation compatible with Machian relation. Brans and Dicke (1961a) reformulated Einstein general relativity with their scalar tensor theory of gravitation, apparently compatible with Mach's inertia Principle. Hoyle and Narlikar (1964,1966) developed a theory of gravitation which was on the Mach's inertia principle and they used waves to communicate gravitational influence between particles. Another form, for Machian relation, called in the literature as Whitrow-Randall relation (Whitrow and Randall,1951) and D. W. Sciama (1953) used electrodynamics' type equations for gravity to extract Machian relation.

But here we want to modify directly the Newtonian dynamics with Mach's inertia principle to show that the Machian Newtonian dynamics is a real alternative for dark matter. We show here that the Milgrom MOND (Milgrom, 1983a., 1983b., 1983c) is an approximate answer and showing that there is no any fundamental constant acceleration a_0 in the physics. Milgrom theorized an observationally equation instead Newtonian force as his model's principal paradigm, claiming a new fundamental constant acceleration a_0 so that

$$a \leq a_0 \Rightarrow g^N = a^2 / a_0$$
$$a > a_0 \Rightarrow g^N = a$$

(2)

So that g^N is Newtonian gravitational intensity and a is acceleration and a_0 is Milgrom constant acceleration. This equation is near to fit with rotation curve of the galaxies. But we can see that

f

the Milgrom formula is inverse engineering of the Tully Fisher relation as it is visible in below proposition that

$$g = a^2 / a_0 \Leftrightarrow v_r^4 = a_0 G M_{<r} \tag{3}$$

Easily we see that the Milgrom gravity formula and Tully-Fisher relation are transferred each other. However, Milgrom formula was rearranged by Poisson type gravity "AQUAL" (Bekenstein and Milgrom 1984) that

$$\nabla \cdot [(g / a_0)\vec{g}] = -4\pi G\rho \tag{4}$$

But clearly this is replacement of the Newtonian g^N in Poisson equation of gravity with g^N in Milgrom's formula. Of course according to the paper by Sanders and McGaugh (2002), one problem with AQUAL (or any scalar-tensor theory in which the scalar field enters as a conformal factor multiplying Einstein's metric) is AQUAL's failure to predict the amount of gravitational lensing actually observed in rich clusters of galaxies. On the other hand, Tully-Fisher (1977) and Faber-Jackson (1976) relations are not accurate in all galaxies suppose fake galaxies deviate from Tully-Fisher and Faber-Jackson relations. Then actually the Milgrom model of modified gravity deviates from the reality in the fake galaxies. Milgrom ever has tried to compensate failures with change of the uncertain parameters similar to the parameter mass to light ratio and also distance uncertainties but there are serious difficulties, especially at the Bullet clusters and galaxies clusters and globular clusters (e.g. Jordi, et al. 2009; Baumgardt, 2006; Sollima and Nipoti, 2009; Aguirre, 2001; Kent, 1987; Gentile, et al. 2011; Famaey, et al. 2007).

TABLE OF CONTENTS

LEGAL NOTES

CHAPTER 1.
MODIFICATION OF NEWTONIAN DYNAMICS WITH MACH INERTIA PRINCIPLE

In Mach's mechanics, the gravitational G is not constant but variable in agreement with Machain relation so that in an N active mass points m_k universe in flat space (frame centre at the rest of the system) around a passive mass point m_p we have

$$\frac{1}{G} = \frac{1}{C^2} \sum_{K=1}^{N} \frac{m_k}{\left|\vec{r}_{kp}\right|} \tag{5}$$

Where $\vec{r}_{kp} = \vec{r}_k - \vec{r}_p$.

This is Machian relation (Einstein-Whitrow-Randall-Sciama-Brans-Dicke relation) and in a continuum shape, Machian relation is written in integral format over the universe that

$$G \frac{\int_U \frac{\rho_r}{|r'-r|} dV}{C^2} = 1 \tag{6}$$

So that r is the spatial position of the passive mass point and r' is where the universe mass points exist and C is the light speed and G is gravitational coefficient at position of passive mass point and ρ is density. Replacement of G in the Newtonian gravity with variable G in Machian relation will generalize it from a local mutual type to a universal version that

$$g^N = \frac{G_N \sum\limits_{k=1}^{N} \frac{m_k}{\left|\vec{r}_{kp}\right|}}{C^2} a \qquad (7)$$

This relation is what the Carl H. Brans (1961b) "Mach's principle and a varying gravitational constant" resulted it by simple dimensional calculus, of course Carl Brans decided to continue on the modified general relativity instead modified Newtonian gravity and also this relation is the same Einstein-Sciama force however Einstein-Sciama force is assumed as an inertial reaction force. Berry (1989) did result this equation by a way that Sciama named it, the law of inertial induction and, a next general way to extract the Machian relation is the way of zero total energy (Berman, 2007, 2008, 2009). Some other scientists also have tried in this way and there are many papers not possible to quote all here. All these ways arrive to an answer named generally "Machian relation". By the way when φ is potential energy, then equation (7) simply is written in the below format that

$$g^N = \frac{\varphi}{C^2} a \qquad (8)$$

We can define the inertia coefficient *in* in the Newtonian law of the inertia as

$$g^N = in \times a \big| in = \varphi / C^2 \qquad (9)$$

The inertia coefficient *in* and G are both coefficients in the law of inertia and then if we assume the inertia coefficient *in* as a variable, the G is constant and inverse. This modified gravity is a scale invariant relation and independent of the meter and second and kilogram, for reality that if k and k' and k'' to be arbitrary coefficients we can show this equation with a general function f

$$f\left(kx, k\,'t, k\,"m\right) = 0 \qquad\qquad (10)$$

And this is the magic of modified gravity by Mach's inertia principle. In the Milgrom MOND too, his formula is scale-invariant at accelerations a $<a_0$, where a_0 corresponds to Milgrom fundamental acceleration (Millgrom, 2009). R. Dicke (1962) also did discuss on the scale invariant gravity in relativistic context as noted by R. Dicke that

"... It is evident that the particular values of the units of mass, length, and time employed are arbitrary and that the laws of physics must be invariant under a general coordinate-dependent transformation of units."

And also scale invariant gravity has been modelled in shape dynamics by Julian Barbour (1982, 2003).

CHAPTER 2.
GENERALIZATION OF MODIFIED GRAVITY IN BOUND LIMITED POTENTIAL ENERGY

2.1 THE BOUND LIMITED GRAVITATIONAL SYSTEMS (QUASI -UNIVERSES)

One of the Mach's principles as listed by Bondi and Samuel (1996) is that

"Mach4: The universe is spatially closed."

As noted by Julian Barbour (2010) that

"However, the Machian view point is only possible if the universe is a closed dynamical system. I shall say something about the possibility of a truly infinite universe at the end of this paper. If we do suppose that the universe is a closed system, we can attempt to describe it by means of a relational configuration space obtained by some quoting with respect to a group of motions."

And as noted by Woodward (2004) that Dennis Sciama adopted a simple statement (Sciama, 1964; Haisch, 2004) that

"Inertial forces are exerted by matter, not by absolute space. In this form the principle contains two ideas:

(1) Inertial forces have a dynamical rather than a kinematical origin, and so must be derived from a field theory [or possibly an action at-a-distance theory in the sense of J.A. Wheeler and R.P. Feynman...]

(2) The whole of the inertial field must be due to sources, so that in solving the inertial field equations the boundary conditions must be chosen appropriately."

In relativistic context too, Machian universe should be a bound system as noted by Amitabha Ghosh (2000) that

"Thus we see here how the Mach principle is entirely intertwined with the theory of general relativity, regarding the logical dependence of the inertial mass of local matter on a closed system."

Then in the Machian frame work, whether observable universe is finite or infinite, the inertial field equations should be solved merely in a definite shape of the universe which we name it here quasi universe.

On the other hand, David Hilbert (2013) famously argued that infinity cannot exist in physical reality. The consequence of this statement is that infinity is not observable and then infinity is meaningless in the Mach's mechanics. Then inertia in the quasi universes doesn't require to integrate on the whole suppose inertia for example in a galaxy as a quasi-universe requires to be integrated over the mass under dominant of the gravitationally bound system. Then the question is that what defines the boundary of inertial field equations?

Mach's mechanics is a fully relational dynamics and then the boundary of the inertial field equations is defined relatively, where the internal inertia is larger than the external inertia means

$$in_{int} \geq in_{ext}$$

(11)

This is resulting that the Machian universe is a finite universe ultimately limited to a universe with light speed at its boundary. Of course we should notice that in relativistic context compatible with especial relativity, the light speed is not possible to occur for mass actually and then the boundary of universe will not appear else at infinity. Then Einstein especial relativity is in contrast with Machian mechanics else when we consider bound limited quasi universes.

2.2 GENERALIZATION OF THE MACHIAN RELATION IN THE QUASI UNIVERSES

By the gravity limitation to the boundaries of the bound limited systems we need to integrate the inertia over the enclosed baryonic mass of the system as association of all mass point particles m_k under dominant of the quasi universe so that from equations (8) and (11), we have in continuum shape that

$$g^N = \frac{G_N \oint_{boundary} \frac{\rho_{r'}}{|\vec{r}'-\vec{r}|}dV}{C^2} a \qquad (12)$$

Then for asymptotical position of the passive mass point we have by shell theorem for a bound limited system that

$$C = v_\infty \qquad (13)$$

Then the coefficient C is not ever light speed suppose generally it is asymptotic speed or speed in horizon of the quasi-universe and just it is light speed at maximum amplitude. Mixing equations (13) and (12) results

$$g^N = \frac{G_N \oint_{boundary} \frac{\rho_{r'}}{|\vec{r}'-\vec{r}|} dV}{v_\infty^2} a \tag{14}$$

This relation is independent of the length and time and mass units showing that the generalized inertial force is also scale invariant and invariant by unit transformations.

We can delete arbitrary Newtonian G_N from equation (14) to obtain a variable G Newtonian type equation $g = a$ so that

$$\frac{G \oint_{boundary} \frac{\rho_{r'}}{|\vec{r}'-\vec{r}|} dV}{v_\infty^2} = 1 \tag{15}$$

This is generalized Machian relation in quasi universes as a strong verification for M-MOND (Machian-MOND).

2.3 THE UNIFICATION OF QUASI UNIVERSES BY UNIVERSAL INTERCONNECTION IN THE MACH'S MECHANICS

Each quasi universe is gravitationally an independent universe that the gravity is defined there with a variable G with a degree of freedom in the rest of the quasi universe (centre of mass invariant by physics laws) and this is compatible with Mach's principle as noted by M. Sachs (2003) that

"The dependence of the inertial mass of localized matter, in particular, on the rest of the matter of the 'universe', is a statement of the Mach principle."

But all quasi universes should follow a unified law for their dependency to their rest of the system by universal interconnection in Machian mechanics and then quasi universal systems in Machian mechanics follow the universe limited to the maximum speed, that is, the light speed at its boundary. This means that Gravitational G in the rest of each quasi universe should be equal with gravitational G in the rest of universe limited to the light speed in its boundary so that

$$G_U(0) = G_{QU}(0) \tag{16}$$

This is compatible with Mach's mechanics for that freedom degree of the gravity is not defined here externally by conceptual absolute elements but fully relational dynamics (mere ordering upon actual objects). Then for central point of the universe we have a central universal G_U so that if we consider arbitrary the frame at the centre of universe, then by equation (6) we obtain

$$G_U \frac{\oint_U \frac{\rho_{r'}}{|\vec{r}'|}}{C^2} = 1 \tag{17}$$

In the next sections we will see that the solar system is positioned presently at the Newtonian regime of the Milky way galaxy and then we have

$$G_U(0) = G_N \tag{18}$$

Then presently, the gravity at the centre of quasi universes is Newtonian but not constant ever.

2.4 AN EQUIVALENT VERSION OF THE MODIFIED GRAVITY BY MACH'S INERTIA PRINCIPLE

In the centre of the quasi universes, the gravity is Newtonian and then from equation (8) we obtain

$$\varphi_o = C^2 \tag{19}$$

Where φ_o means the potential at the center.

The replacement of C, at equation (8) with central potential energy at equation (19) results

$$g^N = \frac{\varphi_r}{\varphi_o} a \tag{20}$$

This equation is possible to be written in the below format too

$$g^N = \frac{\displaystyle\sum_{k=1}^{n} \frac{m_k}{\left|\vec{r}_{kp}\right|}}{\displaystyle\sum_{k=1}^{n} \frac{m_k}{\left|\vec{r}_{ko}\right|}} \tag{21}$$

So that r_{ko} is distance from the centre. Rarely in some area it is possible that φ_r to be larger than φ_o and then in some specific areas it needs to consider fictitious dark matter with negative density, for example in some areas between two very near galaxies as detected in one of the Milgrom paradigms (Milgrom, 1986b) that

- *"A DM interpretation of MOND should give negative density of "dark matter" in some locations (Milgrom 1986b)."*

Also the shape of fictitious dark matter follows the potential energy in the galaxies and φ is not constant in both projected and de-projected density profiles despite the mass integration. Then by fictitious dark matter model it needs ever to use a disk component for imaginary dark matter and this is one of the next paradigms of Milgrom MOND that

- *"Disc galaxies are predicted to exhibit a disc mass discrepancy. In other words, when MOND is interpreted as DM we should deduce a disc component of DM as well as a spheroidal one (Milgrom 1983b,2001; Famaey and McGaugh, 2012)."*

Chapter 3.
Tully-Fisher relation generalization

3.1 Standard and fake galaxies

For a sample of about two dozen elliptical, Robert A. Fish (1964) discovered a relationship between the total potential energy and total mass of the galaxies and, Freeman (1970) commented that Fish's law can also be interpreted more directly to state that central surface densities of the Fish's sample of ellipticals has a universal value. But on the base of the observable mass of the galaxies with different morphologies we observe that the Fish's law is not specified to the elliptical galaxies suppose galaxies all are distributing around the Fish's law so that newly Freeman's law was confirmed for a sample of 30000 Sloan Digital Sky Survey galaxy images (Fathi 2010). Also we may refer to the paper (Allen and Shu 1979) against the selection effect proposed by Disney (1976).

According to the Fish's law, for total mass M of the galaxies in relation with total potential energy we have

$$W = 9.6 \times 10^{-11} M^{3/2} \tag{22}$$

W is associated total potential energy of the system calculated from visible mass.

On the base of the definition of total potential energy, Fish's law is possible to be written in the below form

$$\sum_{j=1}^{N}\sum_{i<j}\frac{Gm_i m_j}{|r_i - r_j|} = 9.6 \times 10^{-11} M^{3/2}$$

(23)

We have a good approximate relation as

$$\sum_{j=1}^{N}\sum_{i<j}\frac{Gm_i m_j}{|r_i - r_j|} = G\sum_{i=1}^{N} m_i \times \sum_{j=1}^{N}\frac{m_j}{|r_j|}$$

(24)

Substituting this relation in the equation (23) results

$$\sum_{j=1}^{N}\frac{m_j}{|r_j|} = \sqrt{2M}$$

(25)

Then we obtain an absolute central scalar potential energy $\dot{\varphi}_o$ as dot-phi that

$$\dot{\varphi}_o = \sqrt{2M}\left|\dot{\varphi}_o = \sum_{j=1}^{N}\frac{m_j}{|r_j|}\right.$$

(26)

The equation (26) is a version of the Fish's law and generally the galaxies are scattering around a standard point that the equation (26) is compatible. We can name this central potential energy of the galaxies as the standard potential energy. But generally, we can consider a scattering coefficient λ' that

$$\dot{\varphi}_o = \lambda'\sqrt{2M}$$

(27)

And as a definition here, a quasi-universe is standard if

$$\lambda' = 1$$

(28)

And if λ' is not unit, the galaxy is defined here the fake.

The observable universe is possible to be considered as a cluster of the super clusters and the universe should follow the universal standard potential energy driven by equation (26). According to the definition of the scalar gravitational potential energy, for center of universe we have for an assumed average density of universe $\bar{\rho}$ that

$$\varphi_o = G_N \int_0^R \frac{\bar{\rho}}{r} dV \tag{29}$$

$$\dot{\varphi}_o = 2\pi\bar{\rho}R^2 \tag{30}$$

If Fish's law is agreement in the observable universe, then equation (30) should be agreement with equation (26). Now if we substitute the equation (30) into the equation (26) we obtain

$$\sqrt{2M} = 2\pi\bar{\rho}R^2 \tag{31}$$

By this relation one results directly universe mean volume density that

$$\bar{\rho}_U = \frac{2}{3\pi R} = 0.042 \times 10^{-26} \tag{32}$$

Early observations verify this size, unlikely with dark matter density. Then Fish's law is agreement with observed baryonic mass distribution of the universe whereas completely disagreement with dark matter density in the universe, theorized almost 25 times larger than baryonic value. As noted by Robert A. Fish (1964) that

"The connection between the potential energy of an elliptical galaxy and the radiation emitted during the contraction of the protogalaxy follows from the Virial theorem."

The gravitational potential energy in Virial theorem never could to discriminate between dark matter and visible matter and then Fish's law never could to be independent of the dark matter density profile. Therefore, full independency of Fish's law from dark matter is showing that there is no dark matter in the galaxies and universe and inverse engineering of the rotation curve of the galaxies used for definition of the dark matter density profiles similar to Navarro-Franklin-White (1996) is not a true way to resolve the gravity puzzle.

3.2 GENERALIZATION OF TULLY-FISHER RELATION BY MODIFIED GRAVITY

By equation (13) and mixing it with equation (19) it is extracted that

$$v_\infty^2 = \varphi_o \tag{33}$$

This equation is generalization of the Tully-Fisher relation in Machian mechanics. By equation (27) embedding into this equation (33) we result a proposition that

$$\dot\varphi_o = \lambda'\sqrt{2M} \xrightarrow{MMOND} v_\infty^4 = \lambda'^2 2G_N^2 M \tag{34}$$

This proposition is a strong validation for M-MOND and a verification for generalization of the Tully-Fisher relation and we see that when we have an assumed standard galaxy ($\lambda'=1$), then it is compatible the Tully-Fisher relation (1977) that $v_\infty^4 = 2G_N^2 M$.

CHAPTER 4.
MILGROM MOND AS AN APPROXIMATION

Consider arbitrary a formula for scalar potential energy in a point at radius r in an isotropic galaxy that

$$\dot{\varphi}_r = \frac{\sqrt{MM_{<r}}}{r} \tag{35}$$

We consider here arbitrary a deviation coefficient λ'' for this relation that

$$\dot{\varphi}_r = \lambda'' \frac{\sqrt{MM_{<r}}}{r} \tag{36}$$

By substituting equation (25) in the equation (21) we obtain for a standard galaxy that

$$g^N = \frac{\dot{\varphi}_r}{\sqrt{2M}} a \tag{37}$$

By substituting equation (36) in the equation (37) we obtain for a standard galaxy that

$$g^N = \lambda''^2 \frac{a_r^2}{2G_N} \tag{38}$$

Milgrom formula is extracted where $\lambda'=1$ and then for a standard galaxy it is extracted below proposition that

$$\dot{\varphi}_r = \lambda'' \frac{\sqrt{MM_{<r}}}{r} \xrightarrow{M-MOND} g_r = \lambda''^2 \frac{a_r^2}{2G_N} \tag{39}$$

This conditional proposition is a next manifest validation for M-MOND and a manifest proposition to show that the Milgrom MOND is an approximate answer specified in a case that the system to be standard and $\lambda''=1$. It is visible that the galaxies and galaxies clusters are scattering around the point that $\lambda''=1$. But for a pure argument we can use from an ideal disk galaxy compatible with Sersic profile as a good approximate density profile here. From shell theorem for scalar potential energy at distance r in an isotropic disk galaxy we have

$$\dot{\varphi}_r = \frac{M_{<r}}{r} + \int_r^R \frac{dm}{r} \tag{40}$$

And then to calculate λ'' we substitute φ_r from equation (36) into this equation

$$M_{<r} + r\int_r^R \frac{dm}{r} = \lambda''\sqrt{MM_r} \tag{41}$$

Writing this equation in the surface density format results

$$\int_0^r \Sigma_r \, rdr + r\int_r^R \Sigma_r \, dr = \lambda''\sqrt{\int_0^R \Sigma_r \, rdr \times \int_0^r \Sigma_r \, rdr} \tag{42}$$

And for a galaxy with Sersic number n=1 and scale length h it is resulted

$$h-(h+r)e^{-r/h} + r\left(e^{-r/h} - e^{-R/h}\right) = \lambda''\sqrt{\left(h-(h+R)e^{-R/h}\right)\left(h-(h+r)e^{-r/h}\right)} \tag{43}$$

this equation is invariant by units' transformation. We can transfer arbitrary the scale of the meter until the length scale h to be 1 and then by equation (43) we obtain

$$\frac{1}{\lambda''} = \frac{\sqrt{1-(1+r)e^{-r}}}{1-e^{-r}} \tag{44}$$

Plotting equation (44) at Figure 1., does show that $\lambda''=1$. Actual density profiles does show better agreement with $\lambda''=1$ because that Sersic profiles are using from an extrapolated central surface density. Then by equation (38) and $\lambda''=1$ we deduce

16

$$g^N = \frac{a_r^2}{2G_N} \tag{45}$$

By comparison this equation with Milgrom formula we obtain that Milgrom has considered a

fundamental constant acceleration a_0, instead $2G_N$ whereas we see that

$$a_0 = 2G_N \tag{46}$$

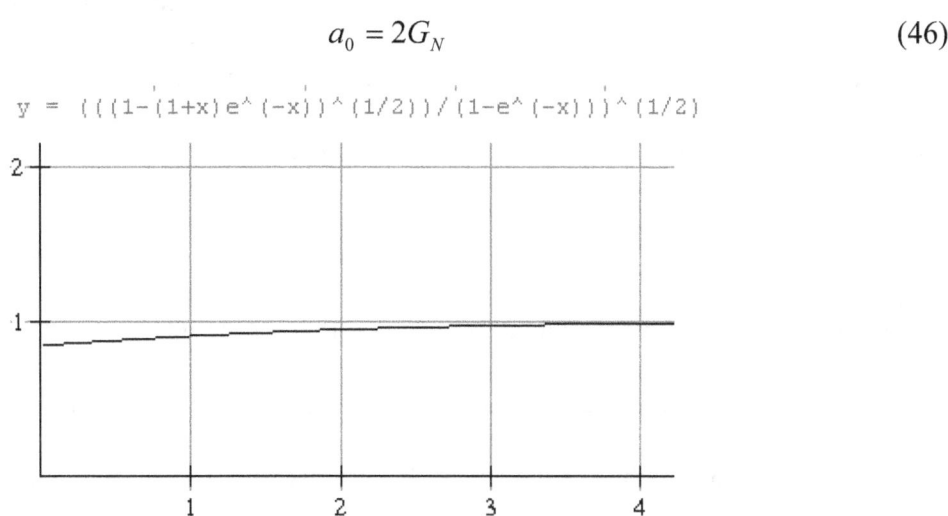

Image 1. Plotting $\lambda''^{-0.5}$ by vertical axis r in the scale of (h=1)

And this is completely agreement with Milgrom phenomenological report for amplitude of his

fundamental acceleration, that is, $a_0 = 1.2 \times 10^{-10}$.

CHAPTER 5.
EXTERNAL FIELD EFFECT AND QUASI NEWTONIAN GRAVITY

By equation (11), while the inertia of an external galaxy A is larger than internal inertia of a host galaxy B, then in the gravity equation of the galaxy B requires to use the inertia in dominant of the external galaxy A. then for a radii r of the galaxy B which the internal inertia is smaller than the inertia of the external galaxy A we have that

$$in_{ext} > in_{int} \Rightarrow g_B = in_A a_B \tag{47}$$

Then critical radii r_c of a host galaxy is radii that the internal inertia is equal with inertia of the external galaxy. For higher radiuses, the gravity should be calculated in the inertia of the external galaxy. For a radii r of the galaxy B with distance L from the external galaxy A, by equations (47) and (37) we have

$$g_B \times \frac{\sqrt{2M_A}}{\dot{\varphi}_{ext}} = a_B \tag{48}$$

And for external inertia we can write as a good approximation that

$$\dot{\varphi}_{ext} = \frac{M_A}{L} \tag{49}$$

And by substituting this relation in the equation (48) we have

$$g_B \times \frac{\sqrt{2G_N}}{\sqrt{G_N M_A / L^2}} = a_B \tag{50}$$

And then

$$g_B \times \frac{2G_N}{g_{ext}} = a_B \Big| g_{ext} = \sqrt{2G_N g_A} \tag{51}$$

This equation (51) results a quasi-Newtonian Gravitational G so that

$$G = G_n \frac{2G_N}{g_{ext}} \tag{52}$$

On the other hand, by equations (11) and (49) and (37) we obtain

$$in = \frac{M/L}{\sqrt{2M}} = \frac{\sqrt{GM/L^2}}{\sqrt{2G}} = \frac{g^M}{2G} \Big| g^M = \sqrt{2G_N g^N} \tag{53}$$

Then by equations (11) and (53), for external field effect it should be agreement

$$2G_N > g_{ext}^M > g_{int}^M \tag{54}$$

And this is the same paradigm reported by Milgrom phenomenology (1983a) that

"An external acceleration field, ge, enters the internal dynamics of a system imbedded in it. For example, if the system's intrinsic acceleration is smaller than ge, and both are smaller than a_0, the internal dynamics is quasi-Newtonian with an effective gravitational constant Ga_0=ge (Milgrom 1983a, 1986a, Bekenstein and Milgrom, 1984). This was applied to various astrophysical systems such as dwarf spheroidal galaxies in the field of a mother galaxy, warp induction by a companion, escape speed from a galaxy, departure from asymptotic flatness of the rotation curve, and others."

CHAPTER 6.
NEWTONIAN REGIME OF HIGH SURFACE BRIGHTNESS GALAXIES AND RELEVANT PARADIGMS

We have a very clear proposition that

$$r_a < r_b \Rightarrow \varphi_a > \varphi_b \qquad (55)$$

This proposition means that in lower radiuses, the scalar potential energy is larger. On the other hand, by equation (20) we find that the gravity in a radii r is Newtonian if

$$\varphi_a = \varphi_b \qquad (56)$$

At the centre the gravity is Newtonian but if there is a maximum size of radius r_t so that

$$\varphi_{r_t} = \varphi_o \qquad (57)$$

Then aligned with the equation (55) it is resulted that

$$r < r_t \Rightarrow g^N = a \qquad (58)$$

And this means that if there is such a radius r_t then we have a Newtonian regime for $r < r_t$. Now question is that is there such a radius at the galaxies which the gravitational potential to be equal with central gravitational potential?

To answer we need to embed the equations (26) and (35) in the equation (56) so that we have

$$\frac{\sqrt{MM_{<r}}}{r} = \sqrt{2M} \tag{59}$$

And then it is deduced that

$$a_t = 2G_N \tag{60}$$

This is showing that the transition radius of a Newtonian regime is at the boundary with acceleration equal to $2G_N$ verifying a next phenomenological paradigm of Milgrom that

- *"In a disc galaxy, whose rotation curve is Vr, that has high central accelerations ($v_r^2/r > a_0$ in the inner regions), the mass discrepancy appears always around the radius where $v_r^2/r = a_0$. In galaxies whose central acceleration is below a_0 (low surface brightness galaxies–LSBs) there should appear a discrepancy at all radii (Milgrom 1983b)."*

Of course λ' and λ'' are not actually exact on the unit and then transition radiuses of the galaxies actually are scattering around a mean value at acceleration equal to $2G_N$.

This phenomenon too has been reported before as one of the Milgrom paradigms that

- *"For a concentrated mass, M, well within its transition radius, $r_t = \sqrt{MG/a_0}$, r_t*

plays a special role (somewhat akin to that of the Schwarzschild radius in General Relativity) since the dynamics changes its behaviour as we cross from smaller to larger radii. For example, a shell of phantom DM may appear around this radius (Milgrom and Sanders, 2008)."

At the paper (McGaugh et al., 2016), McGaugh, Lelli, and Schomberg analysed data from a set of about 150 disk galaxies in the prime. They identified the best-fitting acceleration scale for each of them and found that the distribution is clearly scattering around a mean value at $2G_N$. of course on the same data, a new report (Rodrigues et al., 2018) is showing a monotonically deviation from Milgromian constant which not possible to illustrate aligned with mean value of the Milgrom. probably the reason of discrepancy is technical, on the different styles which have been considered in the analysis, means Gaussian priors by McGaugh et al. instead flat priors over a finite bin.

Also it is manifest that

$$\varphi_r < \varphi_0 \Rightarrow a_r < 2G_N \tag{61}$$

This is showing that deep MOND is appeared at a $< 2G_N$ in agreement with phenomenological reports without any type of cosmological hypothesis versus the Milgrom hypothesis about a fundamental constant acceleration a_0. On the other hand,

we see that $\lambda'' = 1$ and then we have

$$r \leq r_t \Rightarrow \frac{\sqrt{MM_r}}{r} \cong \sqrt{2M} \tag{62}$$

And then it is deduced that

$$r \leq r_t \Rightarrow a_r \cong 2G_N \tag{63}$$

Then in the Newtonian core of the HSB galaxies too, the acceleration will not much exceed than $2G_N$. This phenomenon has been reported before as one of the Milgrom paradigms that

- *"The excess acceleration that MOND produces over Newtonian dynamics, for a given mass distribution, cannot much exceed a_0 (Brada and Milgrom 1998)."*

The surface density and acceleration are at the same dimension and then this paradigm is possible to translate at the surface density form as noted by Milgrom that

- *" Isothermal spheres have mean surface densities $\overline{\Sigma} \leq \Sigma_o = a_o / \pi G$ (Milgrom, 1984) underlying the observed Fish law for quasi-isothermal stellar systems such as elliptical galaxies."*

This is simply possible to be derived from correlation of gravity and mean surface density within transition radius as follows

$$g_t^N = 2G_N \tag{64}$$

And then

$$\frac{M(r_t)}{r_t^2} = \frac{2}{\pi} \tag{65}$$

Then critical mean surface density at transition radii is $\Sigma_t = 2/\pi$.

And generalizable to the mean surface density by the reality visible at equation (63).

Chapter 7.
Galaxies clusters and globular clusters difficulties with Milgrom's MOND

There are serious problems in large scale as noted by Zhao and Famaey (2012) that

"However, a side effect of Bekenstein's exponentially-varying function is that it predicts a value of a_0 of the order of 10^{10000} ms^{-2} once the potential reaches the order of c^2, i.e., a neutron star or a stellar black hole would exhibit an undesirable deep-MOND behaviour."

And the formula by Hodson and Zhao (2017) is an inspired Galaxy clusters show a residual mass discrepancy even when analysed using MOND. It was realized early on (The andWhite, 1988; Gerbal, et al., 1992; Sanders, 1994; Sanders, 1999; Sanders, 2003; Aguirre, et al., 2001; Pointecouteau and Silk, 2005; Takahashi and Chiba, 2007; Angus et al. 2008; Milgrom, 2008) that MOND does not fully explain away the mass discrepancy in galaxy clusters. also when there is external field effect, the reports show deviation from Milgrom formula of gravity (e.g. Derakhshani, 2014) and also external field effect shows different amplitudes of the Milgromian a_0 (e.g. Swaters, 2010) as noted in this paper that

". . . but the average value we find for $a_0 \approx 0.7 \times 10^{-8} cms^{-2}$ is somewhat lower than derived from previous studies. Such lower fitted values of a0 could occur if external gravitational fields are important."

Some theories exist on the subject. Milgrom idea (Milgrom, 2008) is that this matter is baryonic in some yet unidentified form, such as cold, dense clouds. Another possibility that has been considered are massive neutrinos (Sanders, 2003; Angus, 2008).

About the Globular clusters we have the same phenomenon (e.g. Baumgardt, 2006; Ibata, 2011) similar to the cluster of the galaxies included to noticeable deviations from Milgrom MOND.

there are many papers reporting significant deviations from Milgromian constant acceleration not possible to quote all here for example a new result by Frandsen and Petersen (2018) or Ludlow et al. (2017) fit the Milgrom formula to their simulated galaxies, but they find a different value instead Milgromian acceleration a_0. Also the 2006 observation of a pair of colliding galaxy clusters known as the "Bullet Cluster" (Clowe, 2006) poses a significant challenge for all theories proposing a modified gravity solution to the missing mass problem.

But we see here that the sub galaxies in a host galaxy are not totally under dominant of the host galaxy suppose there is a critical radius in each sub galaxy in a cluster that internal inertia is equal to external inertia of mother galaxy. Each mass below this critical radius of sub galaxy is not under dominant of the host galaxy.

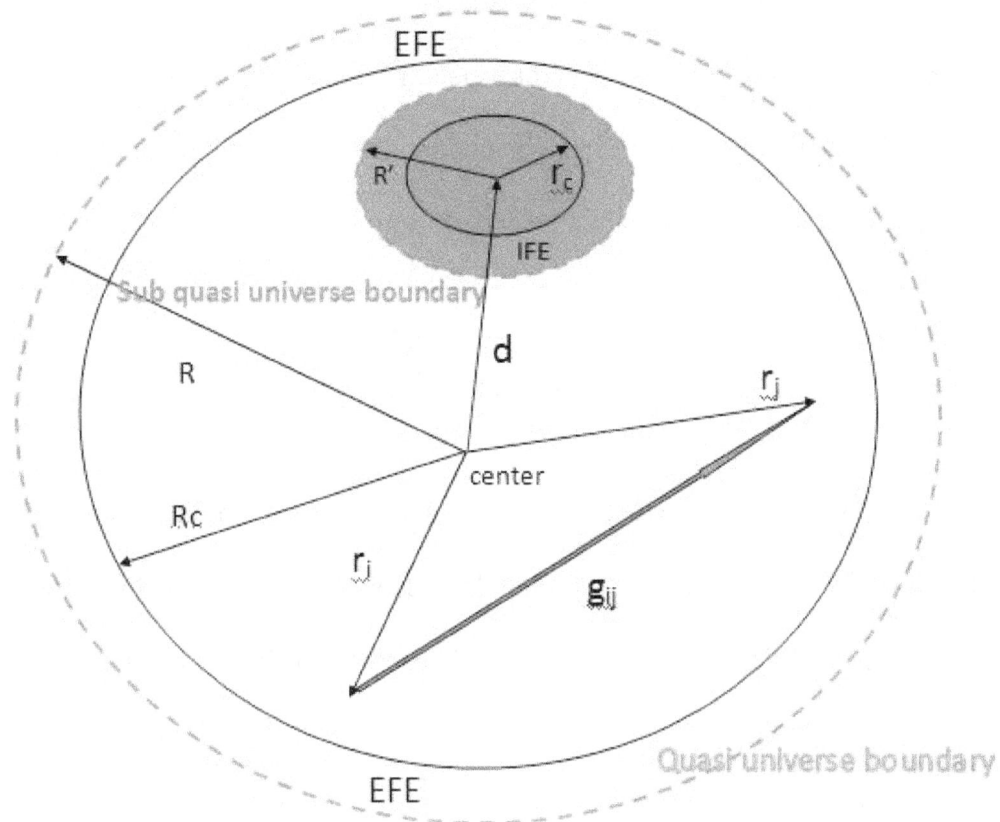

Figure 2. the sub quasi universe in a quasi-universe and generation of EFE by external field effect and IFE by internal field effect

As it is visible at Figure 2, the inertia integration at the point r_j should be integrated over the mass inside the boundary of the quasi universe else where the mass is under dominant of the sub quasi universe ($r < r_c$, when r from center of sub quasi universe) and else, where the mass is under external field effect ($R_c < r < R$, when r is from center of quasi universe). this internal field effect is discounting the masses under dominant of the sub galaxies from potential energy integration in the host galaxy. Then the mass under dominant of the dwarf galaxies and globular clusters inside a host galaxy or galaxies in the galaxies clusters or the masses of the host galaxy under dominant of the external galaxies are not used in the inertia integration. Mathematically when there are sub quasi universes in a quasi-universe, from equation (11) it needs to discount the mass under dominant of the sub quasi universes from the inertia integration. Then equation (37) under the EFE and IFE are written in the general form that

$$g^N = \frac{\sum_k \frac{m_k}{|r_k|} - \sum_{k \in EPE} \frac{m_k}{|r_k|} - \sum_{k \in IPE} \frac{m_k}{|r_k|}}{\sqrt{2M}} a$$

(66)

Manifestly this is resulting fictitious missing mass even in the context of the Milgrom MOND as reported in the literature and correlation of the deviation amplitude to a part of potential energy under dominant of the internal and external fields is visible here. Equation (66) is possible to write in the below face that

$$g^N = \frac{\sum_k \frac{m_k}{|r_k|} - \sum_{k \in EFE} \frac{m_k}{|r_k|} - \sum_{k \in IFE} \frac{m_k}{|r_k|}}{\sqrt{2M}} \times \frac{a^2}{a}$$

(67)

Then we obtain a fictitious variable A_0 as a generalization for Milgrom constant acceleration as

$$a = \sqrt{A_0 g^N}$$

(68)

Then from equation (67) we have

$$A_0 = \sqrt{2a_o G_N} \times \frac{\sqrt{MM_{<r}}}{r} \times \frac{1}{\sum_k \frac{m_k}{|r_k|} - \sum_{k \in EFE} \frac{m_k}{|r_k|} - \sum_{k \in IFE} \frac{m_k}{|r_k|}}$$

(69)

28

$$A_0 = \frac{a_0}{\left(1 - \dfrac{\varphi_{EFE+IFE}}{\varphi_{TOT}}\right)^2} \qquad (70)$$

The dimension of the A_0 is at the same a_0. In galaxies clusters we have external potential effect of the cluster on the individual galaxies inside the cluster for example an effect has been assumed from Virgo cluster on theMilky way (Famaey, 2007). then we should discount the internal potential effect (IPE) means

$$A_0 = \frac{a_0}{\left(1 - \dfrac{\varphi_{IFE}}{\varphi_{TOT}}\right)^2} \qquad (71)$$

Using the equation (11) to realize the matter under dominant of the galaxies inside a cluster we see that the equation (71) is very agreement with observations of galaxies clusters (e. g. Hodson and Zhao, 2017) so that when the half of the baryonic mass is under the IPF, then the fictitious function of A_0 will increase to $4a_0$. Equation (71) is showing mathematically an infinity when $\varphi_{IFE}/\varphi = 1$. But actually it is clear that it is hard to find even a galaxy cluster with $\varphi_{IFE}/\varphi = 0.75$ and then actually we need to consider a maximum in agreement with results in (Hodson and Zhao, 2017).

Bekenstein (2011) has proposed an exponential increase of A_0 by potential energy and Zhao and Famaey (2018) has proposed an EMOND mechanism to resolve the puzzle of clusters discrepancy and such a way has been continued by Hodson and Zhao (2017). But Bekenstein formula encounters withformula as noted by Hodson and Zhao that:

"For this approach, we are inspired by an alternate modification to gravity that uses a density dependent modification to gravity to try and explain galaxies and galaxy clusters without the need for dark matter (Matsakos and Diaferio 2016)."

On the other hand, for a system similar to a Globular cluster we have external potential of the mother galaxy and then in the globular galaxies, the equation (70) results again discrepancy with Milgrom MOND so that

$$A_0 = \frac{a_0}{\left(1 - \frac{\varphi_{EFE}}{\varphi_{TOT}}\right)^2} \tag{72}$$

We can see that how much the external potential effect of the mother galaxy is rather in proportion to the total potential energy then we have larger discrepancy and A_0 is increased rather.

CHAPTER 8.

BULLET CLUSTERS AND MULTI NUCLEI GALAXIES DIFFICULTIES WITH MILGROM'S MOND

For bullet galaxies or multi-nuclei galaxies we can see that these galaxies are irregular $\lambda''=1$ and non-standard λ' and so, the Milgrom MOND is not match in these galaxies so that we see high discrepancies for bullet cluster 1E0657-56 (Clowe, et al., 2006), Or as a new paper by Israa, et al. (2018) we see that in central region of the NGC 3256 which the gravity should be Newtonian on the context of the Milgrom MOND, we have a high discrepancy with Milgrom MOND. For NGC 3256 we use from equation (20) in general shape under the effect of the external potential that

$$g^N = \frac{\dot{\varphi}_r}{\dot{\varphi}_o} a \tag{73}$$

Above the radius 9.2 arcsec, the mass is under the external Field effect and this is deduced from the reality that the total mass profile breaks (Figure 3.) in this radii very sharp to a Quasi-Newtonian type gravity (Milgrom, 1983a, 1986b; Bekenstein and Milgrom, 1984). Then by shell theorem for potential energy we have for a radius r = 9.2 of the NGC 3256 that

$$g^N = \frac{\dfrac{M_r}{r} + \displaystyle\int_r^{9.2} \dfrac{dm}{r}}{\displaystyle\int_0^{9.2} \dfrac{dm}{r}} a \tag{74}$$

On the base of the baryonic mass profile (Israa, et al., 2018) we consider the radius 0.8 arcsec as

a minimum radius which the mass is readable in confidence so that

$$\dot{\varphi}_0 = \int_0^{9.2} \frac{dm}{r} = \int_0^{0.8} \frac{dm}{r} + \int_{0.8}^{9.2} \frac{dm}{r} \tag{75}$$

and referring to paper by Israa, et al. (2018), we see that

$$m = 1.87 \times 10^{20} r + 3.21 \times 10^{39} \big| (0.8 arc\,sec) \leq r \leq (9.2 arc\,sec) \tag{76}$$

By baryonic matter data from (Israa, et al. 2018) we have $\dot{\varphi}_0 > 2 \times 10^{21}$ but problem is that there

is doubt about the light to mass ratio below the 0.8 arcsec for that the radius 0.8 arcsec is a point

that the gravity is Newtonian on the Baryonic matter and then, on the base of M-MOND we

should have a Newtonian regime below the 0.8

arcsec. Then for r<0.8 it is suitable to use from mass profile deduced by total mass generated by

Newtonian gravity assumption. we have a good approximation for this mass profile as

$$r \leq 0.8 \Rightarrow m = 200 r^2 \tag{77}$$

then by equations (75) and (76) and (77) we obtain that

$$\dot{\varphi}_o \equiv 21.6 \times 10^{20} \tag{78}$$

then calculation of the central potential energy by total mass detected by Newtonian gravity for

r < 0.8 is too showing a near answer to its calculation by baryonic matter profile at (Israa, et al.,

2018) and then this size of central potential energy is on the confidence, whether we use Newtonian gravity for r <0.8 or using baryonic matter by (Israa, et al., 2018). By the way we obtain for NGC 3256 that

$$\lambda' = \frac{\dot{\varphi}_o}{\sqrt{2M}} = \frac{21.6 \times 10^{20}}{\sqrt{2 \times 1.6 \times 10^{40}}} = 12 \tag{79}$$

And then NGC 3256 is a highly fake galaxy. The equation (74) is written as

$$g_r^N = \frac{\frac{M_{<r}}{r} + \int_r^{9.2} \frac{dm}{r}}{21.6 \times 10^{20}} a_r \tag{80}$$

Substituting equation (76) in this equation results (r in arcsec) that

$$g_r^N = \left(0.086 + \frac{0.291}{r} + 0.086 \ln\left(\frac{9.2}{r}\right) \right) \times a_r \tag{81}$$

This is carefully agreement with reported accelerations by Israa, et al. (2018) and also we see that for NGC 3256 at radius 1.7 kpc we observe that a = 8g^N versus the Milgrom MOND which predicts Newtonian gravity at this radius.

Of course we should notice that the $\dot{\varphi}_r$ near the centre should near to $\dot{\varphi}_o$ but NGC 3256 in reality is not a galaxy suppose it is two galaxy colliding together reasonable for a negative value of fictitious dark matter for radiuses lower than the distance between their nuclei. Reason is that in such a colliding galaxies, the centre of mass is not exactly at the same point that the potential energy is maximum and such a negative density of fictitious dark matter in the regions between the nearby galaxies has been reported by Milgrom (1986a). But this negative dark matter is neglect able here. in reality we can see that we need to modify a bit the baryonic mass profile at (Israa, et al., 2018) and this is done by green line in the Figure 3.because of the inflation of light L in this Newtonian regime. in reality NGC 3256 is not just a galaxy suppose it is included to the two colliding galaxies. then we need to consider high values of the frictional radiation deduced by collision of these galaxies. this effect increases highly the light to mass ratio in the dense areas of the NGC 3256 as a source for inflation of visible light below the 0.8 arcsec.

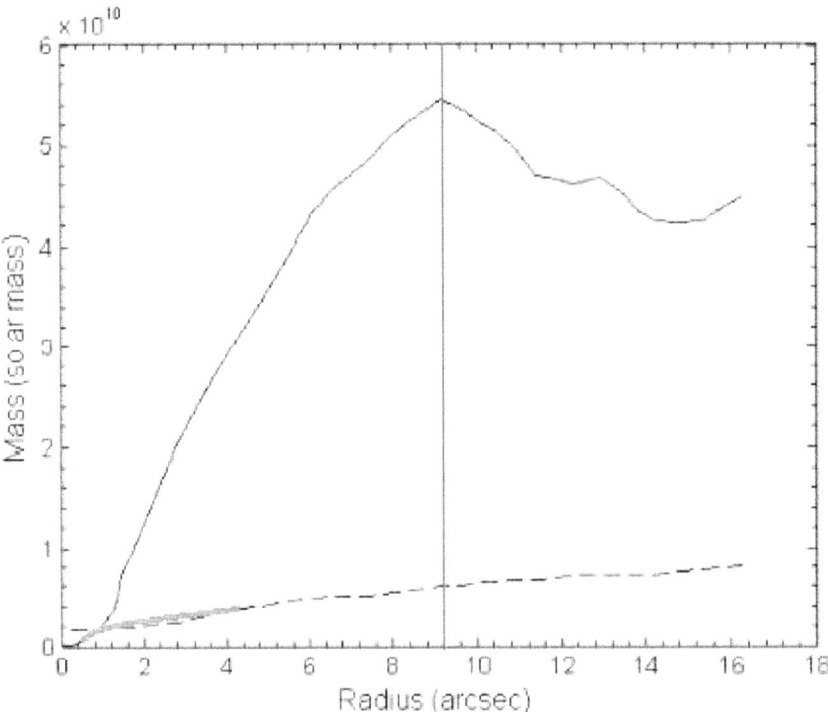

Fig. 3 – Total mass and baryonic mass of NGC 3256 as a function of radius by Israa et al. The solid line is the total mass of NGC 3256 and the dashed line is the baryonic component of NGC3256 and very sharp break of the gravity gradient at r=9.2 arcsec.

On the other hand, from total mass profile at Figure 3, we see a very sharp break at the gravity gradient at the point 9.2 in arcsec so that changing it to a quasi-Newtonian type for radiuses rather than 9.2 arcsec. Quasi Newtonian type gravity is just when we have an EFE on the host galaxy NGC 3256. Then according to the equations (11) and (20), the internal inertia and external inertia at 9.2 arcsec should be equal so that

$$in_{ext} = \frac{M_{<9.2}}{\dot{\varphi}_o} = 0.12 \tag{82}$$

This external filed effect should be relevant to a nearby galaxy for the NGC 3256. If we assume such a galaxy with mass M' and distance L' from the NGC 3256 then we have from equations (37) and (82) that

$$in_{ext} = \frac{\frac{M'}{L'}}{\sqrt{2M'}} = \sqrt{\frac{M'}{2L'^2}} = 0.12 \tag{83}$$

$$g_{ext}^{N} = 2 \times 10^{-12} \tag{84}$$

All galaxies in the universe have smaller effect and greatest effect is just relevant to the nearby galaxy NGC 3256C. It is wonderful that NGC 3256C has a value of g on the NGC 3256 equal to 2×10^{-12}. Then for NGC 3256, in a similar mathematical way used in equation (53) we have

$$r \geq 9.2 \Rightarrow g_{r}^{N} = \frac{\sqrt{2G_{N} g_{(NGC-3256C)}^{N}}}{2G_{N}} - a_{r} \tag{85}$$

And this equation shows a quasi-Newtonian gravity which causes to decrease rapidly the acceleration for radiuses larger than 9.2 arcsec in agreement with rotation curve detected by

35

Israa, et al. (2018). it is interest that on the base of the Milgrom MOND (Milgrom, 1983a), it is impossible such an EFE because that at radius 9.2 arcsec, rotation curves in Milgrom MOND context are showing that $a_{9.2} = 2.8 \times 2G_N$. Now question is that which one of the galaxies are agreement with high discrepancy with Milgrom MOND similar to the NGC 3256?

We observe that the high discrepancy is appeared when the galaxy is highly fake and in realty we have a general condition for "Israael effect" (Israa et al. effect) that

$$\forall r > \varepsilon \left(\varepsilon \to 0 \right) \Rightarrow \int_0^r \frac{dm}{r} > \int_r^R \frac{dm}{r} \tag{86}$$

How much the domain of the radiuses agreement in this proposition is wider and nearer to the centre we have a larger λ' and the galaxy is the more fake.

CHAPTER 9.
MODIFIED GRAVITY IN THE SOLAR SYSTEM

On the base of the above results, the general equation of gravity in a point p of Milky Way galaxy is Newtonian shape with variable G limited to the boundary of the Milky way galaxy. Presently by M-MOND we have

$$G = G_N \frac{\dot{\varphi}(o)}{\dot{\varphi}(\vec{r}_p)} \tag{87}$$

Then by equation (27) we have

$$G = G_N \times \frac{\lambda' \sqrt{2M}}{\sum\limits_{k=1}^{N} \frac{m_k}{|\vec{r}_{kp}|}} \tag{88}$$

M is total mass of the Milky way galaxy and n should be the number of the baryonic particles M_k limited to the boundary of the Milky way galaxy. By substituting equation (35) in the equation (88) we obtain

$$G = G_N \times \frac{\lambda' r}{\sqrt{M_{<r}/2}} \tag{89}$$

In the solar system which the gravity is Newtonian and then by equation (89) we have

$$\frac{\lambda' R}{\sqrt{M_{<R}/2}} = 1 \tag{90}$$

So that R is distance of Earth from center of the galaxy and by this equation it is resulted that

$$g_R = \lambda'^2 \times 2G_N \qquad (91)$$

And by comparing the equation (60) with equation (91) and assuming Milky Way as a standard galaxy we obtain

$$a_t = g_R \qquad (92)$$

This equation is showing that the solar system should be on the exact point of the transition radii r_t as the boundary of the Newtonian core of the Milky Way galaxy and this is true and Milky Way is verifying this reality as a verification for newton gravity modified by Mach's principle here.

CHAPTER 10.
FLATTENING OF THE VELOCITY DISPERSION

Recent velocity measurements for several hundred stars per dSph demonstrates that dsph velocity dispersion remain approximately at with radius. For example, in (Walker et al. 2007) it has been presented velocity dispersion profiles for seven dwarf satellites of the Milky Way Carina, Draco, Fornax, Leo I, Leo II, Sculptor, and Sextans and all the measured dSphs exhibit approximately flat velocity dispersion profiles. Theoretical answer for this flattening of the velocity dispersion in these dwarf galaxies has been on the assumption that dSph are equilibrium systems embedded within dark matter halos however as assessed by (Walker et al. 2007), the stellar velocity distributions are highly anisotropic or ongoing tidal disruption invalidates the assumption of the equilibrium (e.g. kroupa 1997).

One of the Milgrom paradigms is that

"For spheroidal systems a mass-velocity-dispersion relation $\sigma^4 = a_0 G_N M$ is predicted under some circumstances. According to MOND, this is the fact underlying the observed Faber-Jackson relation for elliptical galaxies, which are approximately isothermal spheres (Milgrom 1984). For instance, this relation holds approximately for all isothermal spheres having a constant velocity dispersion and constant velocity anisotropy ratio (Milgrom 1984). in the deep MOND regime, of the form $(4/9)a_0 G_N M$, where σ is the 3-D rms velocity dispersion."

Milgrom MOND is a good approximation from MMOND and then using from Milgrom formula for gravity in deep MOND will result a near answer. Milgrom has used from complicated format to argue his paradigm but we can use a simple way to argue. Velocity dispersion of relevant galaxies is possible to calculate by Virial theorem that

$$\varphi_{tot} = M_{tot} \bar{\sigma}^2 \tag{93}$$

As a relation between total potential energy and mean velocity dispersion of the system. The total potential energy in extended form is written as

$$\bar{\varphi}_{tot} = \int \rho (g \cdot r) dV \tag{94}$$

When we use Newtonian gravitational intensity gN 696 in this equation, then deduced velocity dispersion is not agreement with observations. Milgrom formula is a good approximation from Machian modified gravity and then it is suitable to use Milgrom formula of the gravity here. Milgrom (1984, 1994) has used complicated calculations to extract velocity dispersion of the isolated isothermal spheres. But we can extract total potential energy by Milgrom gravity formula simply as

$$\bar{\varphi}_{tot} = 4\pi \int \rho \sqrt{a_0 g} r^3 dr \tag{95}$$

For isothermal spheres we have $\rho = kr^{-2}$ and then substituting this density profile at equation (95) we result

$$\bar{\varphi}_{tot} = 4\pi k \sqrt{a_0 G} \int \sqrt{M_{<r}} dr \tag{96}$$

For isothermal sphere we have $M_{<r} = 4\pi kr$ and then substituting this equation at equation (96) results

$$\overline{\varphi}_{tot} = \sqrt{a_0 G} \times \frac{2}{3} M^{3/2} \qquad (97)$$

By Virial theorem as a correlation between total potential energy and kinetic energy $\varphi_{tot} = 2T$

and mixing with relation $T = \frac{1}{2}\sigma^2$ we deduce from equation (97) that

$$\frac{2}{3}\sqrt{a_0 G_N M} = \sigma^2 \qquad (98)$$

And for different galaxies with different density profiles it was found (Milgrom,

1984., Sanders, 2010), $\frac{4}{9} \leq \frac{\sigma^4}{a_0 G_N M} \leq 1$.

Of course Milgrom MOND is not agreement perfectly with galaxies velocity dispersions and

there are difficulties with Milgrom MOND as mentioned above. Flattening of the velocity

dispersion in the galaxies is extracted in similar way with rotation curve flattening when we

have a similar relation for velocity dispersion in relation with gravity.

References

Angus G. W., B. Famaey, B. A. Buote, Mon. Not. R. Astr. Soc. 387, 1470 (2008).

Allen R. J., and F. H. Shu, Astronomical Journal. 277, 67 (1979).

Alistair O. Hodson1 and Hongsheng Zhao1, Astronomy Astrophysics manuscript Vol. 598, 127 (2017).

Aguirre A., J. Schaye, E. Quataert, The Astrophysical Journal. 561, 550 (2001).

Brans C., and R. H. Dicke, Phys. Rev. 124, 925 (1961a).

Brans C., Mach's Principle and a Varying Gravitational Constant. unpublished Ph.D. Thesis, Princeton University. (1961b).

Berry M. V., Principles of Cosmology and Gravitation, Adam Hilger, Bristol. (1989).

Berman M. S, Astrophysics and Space Science. 312, 275 (2007).

Berman M. S., Astrophysics and Space Science. 314, 319 (2008).

Berman M. S., International Journal of Theoretical Physics. 48, 3278 (2009).

Brada R., and M. Milgrom, Astrophysical Journal. 519, 590 (1998).

Begeman k. G., A. H. Broeils, R. H. Sanders, MNRAS 249, 523 (1991).

Bekenstein j., Seminaires de l'IAP, (2011) [www.iap.fr/activites/seminaires/IAP/2011/ Presentation737

s/bekenstein.pdf] Bondi H., and J. Samuel, Physics Letters A. 2228, 121 (1996).

Barbour J., H. Pfister,Mach's Principle From Newton's Bucket to Quantum Gravity, Birkhauser, Boston (1995).

Barbour J., Proc. R. Soc. Lond. A 382, 295 (1982).

Barbour J., Classical and Quantum Gravity. 20, 1543 (2003).

Barbour J., Foundations of Physics. 40, 1263 (2010).

Baumgardt H., Testing MOND with globular clusters. 21st IAP Colloquium: Mass Profiles and Shapes of Cosmological Structures . Parc d'Activites de Courtaboeuf, Les Ulis, France: EDP Sciences. (2006).

Bekenstein J., and M. Milgrom, Astrophys. J. 286, 7 (1984).

Clowe D., M. Bradac˘, A. H. Gonzalez, M. Markevitch, S. W. Randall, C. Jones, D. Zaritsky, The Astrophysical Journal Letters. 648, 109 (2006).

Dicke R. H., Phys. Rev. 125, 2163 (1962).

Derakhshani K., The Astrophysical Journal. 783, 48 (2014).

Disney M. J., Nature. 263, 573 (1976).

Einstein A., Vierteljahrsschrift fur GerichtlicheMedizin und Offentliches Sanitatswesen, 44, 37 (1912).

Einstein A., Prinzipielles zur allgemeinen Relativitatstheorie. Annalen der Physik, 55:241 (1918).

Einstein A., The meaning of relativity. Princeton, fifth edition, p. 99-108 (1959).

Faber S. M., and R. E. Jackson, Apj. 204, 668 (1976).

Famaey B., and S. S. McGaugh, Living Reviews in Relativity. 15, 159 (2012).

Fish R. A., Ap. J. 139, 284 (1964).

Famaey B., G. Gentile, J. -p. Bruneton, H. Zhao, phys. Rev. D. 75, 063002 (2007).

Freeman K. C., Astrophysics. J. 160, 811 (1970).

Fathi K., Astrophysical Journal. 722, 120 (2010).

Frandsen T. M., and J. Petersen, Investigating Dark Matter and MOND Models with Galactic Rotation Curve Data. (arXiv:1805.10706v1).

Ghosh A., Origin of inertia, the extended Mach's principle and cosmological consequences. Aperion (2000).

Gentile G., B. Famaey, W. J. G. de Blok, Astron. Astro-phys. 527, 76 (2011).

Gerbal D., F. Durret, M. Lachieze-Rey, G. Lima-Neto, Astron. Astrophys. 262, 395 (1992).

Hilbert D., in David Hilbert's Lectures on the Foundations of Arithmetics and Logic 19171933. (Springer, Heidelberg, 2013).

Hoyle F., and J. V. Narlikar, Proceedings of the Royal Society of London A, 282, 191 (1964).

Hoyle F., and J. V. Narlikar, Proceedings of the Royal Society of London A, 294, 138 (1966).

Ibata R., A. Sollima, C. Nipoti, M. Bellazzini, S. C. Chapman, E. Dalessandro, The Astrophysical Journal. 738, 186 (2011).

Israa Abdulqasim Mohammed Ali, Chorng-Yuan Hwang, Zamri Zainal Abidin, Adele Laurie Plunkett, Sains Malaysiana 47(6): 1241 (2018).

Jordi K., E. K. Grebel, M. Hilker, H. Baumgardt, M. Frank, P. Kroupa, H. Haghi, P. Cot, S. G. Djorgovski, AJ. 137, 4586 (2009).

Kent S. M., AJ. 93, 816 (1987).

Kroupa P., New Astronomy, 2, 139 (1997).

Ludlow A. D., A. Ben´ıtez-Llambay, M. Schaller, et al., Physical Review Letters. 118, 161103 (2017).

Mach E., The science of mechanics, Open Court, 1960.

Milgrom M., Solutions for the modified Newtonian dynamics field equation. ApJ, 302, 617 (1986a).

Milgrom M., and R. H. Sanders, Astronomical Journal. 678, 131 (2008).

McGaugh S. S., F. Lelli, J. M. Schombert, Physical Review Letters. 117, 201101 (2016).

Milgrom M., Astrophys. J. 270, 365 (1983a).

Milgrom M., Astrophys. J. 270, 371 (1983b).

Milgrom M., Astrophys. J. 270, 384 (1983c).

Milgrom M., APJ. 698, 1630 (2009).

Milgrom M., APJ. 306, 9 (1986b).

Milgrom M., Mon. Not. R. Astron. Soc. 326, 1261 (2001).

Milgrom M., Astrophysical journal. 287, 571 (1984).

Milgrom M., Ann. Phys. 229, 384 (1994).

Milgrom M., New Astron. Rev. 51, 906 (2008).

McGaugh S. S., Canadian Journal of Physics. 93 (2): 250–259 (2014).

Navarro J. F., C. S. Frenk, S.D.M. White, The Astrophysical Journal. 462: 563. (1996).

Pointecouteau E., and J. Silk, Mon. Not. R. Astr. Soc. 364, 654 (2005).

Rodrigues D. C., V. Marra, A. del Popolo, Z. Davari, Nature Astronomy. 2, 668 (2018).

Sanders R. H., Astron. Astrophys. 284, 31 (1994).

Sanders R. H., Astrophys. J. 512, 23 (1999).

Sanders R. H., Mon. Not. R. Astr. Soc. 342, 901 (2003).

Sciama D., Rev. Mod. Phys. 36, 463 (1964).

Sachs M., Mach's Principle and the Origin of Inertia. Aperion. 2003.

Sollima A., and C. Nipoti, Mon. Not. R. Astron. Soc. 000, 1 (2009).

Takahashi R., and T. Chiba, . Astrophys. J. 671, 45 (2007).

Tully R. B., and J. R. Fisher, Astr. Ap. 54, 661 (1977).

The L. H., and S. D. M. White, Astronomical Journal. 95, 1642 (1988).

Swaters R. A., R. H. Sanders, S. S. McGaugh, The Astrophysical Journal. 718, 380 (2010).

Sanders R. H., Mon. Not. R. Astr. Soc. 407, 1128 (2010).

Sanders, Robert H.; McGaugh, Stacy S. (2002). "Modified Newtonian dynamics as an alternative to dark matter". Annual Review of Astronomy and Astrophysics. 40 (1): 263–317.

Walker M. G., M. Mateo, E. W. Olszewski, O. Y. Gnedin, X. Wang, B. Sen, M. Woodroofe, The Astrophysical Journal. 667, 53 (2007).

Woodward J. F., Foundations of Physics. 34, 1575 (2004).

Whitrow G.,and D. Randall, MN-RAS,111, 455 (1951).

Zhao H., B. Famaey, Physical Review D. Vol. 86, id. 067301 (2012).

ABOUT THE AUTHOR

MOHSEN LUTEPHY

visit (https://www.researchgate.net/profile/Mohsen_Lutephy)

Email address: lutephy@gmail.com

OTHER BOOKS BY (AUTHOR)

List your other kindle books with a link to the page

https://www.amazon.com/MoED-modification-electro-dynamics-priciple/dp/1537165690

https://www.amazon.co.uk/absolute-physics-non-scale-mechanics/dp/1492164216

https://www.amazon.com/absolute-dynamics-fundamentals-new-paradigm/dp/1540583945

CAN I ASK A FAVOUR? (USE A HEADING THAT WONT SHOW IN TOC)

If you enjoyed this book, found it useful or otherwise then I'd really appreciate it if you would post a short review on Amazon. I do read all the reviews personally so that I can continually write what people are wanting.

If you'd like to leave a review then please visit the link below:

(add the link to your kindle book. You will need to wait until the book is published before being able to get this link)

Thanks for your support!

www.ingramcontent.com/pod-product-compliance
Lightning Source LLC
Chambersburg PA
CBHW080606180526
45168CB00007B/2796